CBD Oil for Anxiety

Effective Treatment Of Chronic Pain, Depression, Children Anxiety, Arthritis, Cancer And ADHD

Gabriella Brown

Table of Contents

CBD OIL FOR ANXIETY 1

FREE BONUS 4

INTRODUCTION 6

CHAPTER 1 8

    WHAT YOU OUGHT TO FIND OUT ABOUT CBD OIL FOR ANXIETY 8

CHAPTER 2 15

    WHAT'S CBD? 15

        CBD Essential oil Effects 16

        Cbd Unwanted Effects 20

        How Do Cbd Oil Assist With Anxiety? 22

        Neurochemistry and CBD 23

CHAPTER 3 28

    CBD ESSENTIAL OIL FOR CHILDREN WITH ANXIETY 28

CHAPTER 4 36

    USING CBD ESSENTIAL OIL FOR TREATING ANXIETY 36
    USES 37

CHAPTER 5 43

    SAFETY 43

        7 Natural Remedies to lessen Anxiety 45

        CBD Essential Oil And Anxiety 53

        Risks 57

CHAPTER 6 60

    MY EXPERIENCE USING CBD ESSENTIAL OIL FOR KIDS AND ADULTS WITH ANXIETY 60

CHAPTER 7 65

How CBD Essential Oil For Kids And Adults Works *65*
Choosing the proper CBD Essential oil for Anxiety *71*
ACKNOWLEDGMENTS **77**

FREE BONUS

Get new and free books by joining Engolee Book Club www.engolee.com/bookdeals

Join Other Subscribers Here

You can visit and join Engolee Book website at www.engolee.com/bookdeals for update, questions as well as to contact us.

Copyright © 2019 by Gabriella Brown

All rights reserved. No part of this publication may be reproduced, distributed, or transmitted in any form or by any means, including photocopying, recording, or other electronic or mechanical methods, without the prior written permission of the publisher, except in the case of brief quotations embodied in critical reviews and certain other non-commercial uses permitted by copyright law.

INTRODUCTION

Lately, cannabidiol (CBD) oil has turned into a widely favored treatment for anxiety. Although some individuals take CBD essential oil to soothe their everyday concerns, others utilize it to treat much more serious conditions like generalized panic.

Unlike tetrahydrocannabinol (or THC, another chemical substance within marijuana), cannabidiol doesn't create a "high" when consumed.

Are you worried that your anxious child may need medication? Well, stress in children is a significant problem, you aren't alone. Regarding to psychom, panic in children can be considered a stage in children's development as they start to connect to the world.

Cannabidiol (CBD) oil is quickly becoming one of the

biggest trends in beauty and wellness, as top athletes, celebrities, and doctors embrace its anti-anxiety, antioxidant, and pain-relieving properties. A powerful anti-inflammatory agent, this non-psychoactive compound of the cannabis plant can benefit the body and skin in myriad ways.

Chapter 1

What You Ought To Find Out About Cbd Oil For Anxiety

During the last decade, a growing number of individuals have started using CBD oil for anxiety and stress alleviation.

With the passing of the 2018 farm bill in America and the entire legalization of commercial hemp, CBD is defined to become nation-wide home treatment for anxiety, depression, and a great many other debilitating conditions.

Despite the fact that research continues to be in the first stages in establishing whether hemp oil, cannabis oil or other marijuana products represent a viable treatment for anxiety, depression, and stress, people who've used them state an overall reduction in symptoms.

Here we make an effort to clear up some typically common misconceptions about CBD, how to locate it, how to utilize it, and how it can benefit you with anxiety.

To understand the consequences and health advantages of CBD, cannabis oil, hemp oil, and medical cannabis, we can look at how these supplements affect our brain and therefore our body.

What's Anxiety?

Stress is among the most typical emotional and behavioral issues that our culture is currently working with. Actually, a 2017 statement released by the World Health Business suggests almost 264 million people live with stress worldwide.

Obviously, researchers and health care professionals

from all over the world are in a continuous seek out for new ways to help people manage anxiety and stress.

The bottom line is, anxiety is a disorder seen as a restlessness and excessive of being concerned. Those folks who are coping with it have trouble making decisions, speaking in public areas, tolerating criticism, or requesting our manager for an increase.

From social panic and anxiety disorder to post-traumatic stress disorder (PTSD) and agoraphobia, anxiety will come in many 'designs and sizes.'

But no matter the kind of anxiety you may be working with, a dosage of cannabidiol will help you deal with the unpleasant symptoms associated with this problem.

What Are The Original Treatments For Anxiety

The first recorded use of cannabis for anxiety alleviation was around the entire year 1500 in India. Since then, experts and health care experts have been occupied studying the consequences of cannabinoids as a potential treatment for an array of brain diseases and psychological problems like depressive disorder, post-traumatic stress, work-related stress, persistent pain, Alzheimer's disease, and so many more.

Therapy for Anxiety

Probably one of the most popular treatment plans for stress is therapy. This calls for a seat with a certified counselor or psychotherapist who can help you understand the primary of your trouble and take the first steps toward curing.

Through the use of techniques such as publicity therapy, thought challenging, or Socratic questioning, you can overcome the irrational thinking patterns that gas your continuous worrying and gain the courage to live an anxiety-free life.

Medications for Anxiety

Medications have been with us for decades, as well as for reasonable. When your panic is so intense, you can't even escape the home or consider yourself, therapy only is simply not enough.

Luckily, the pharmaceutical industry has produced an array of medications for anxiety, known as anxiolytics. But exactly like some other psychiatric medications, anxiolytics can have numerous unpleasant part effects.

Because of this, some healthcare specialists have turned their attention towards cannabinoids from hemp or weed. Actually, numerous experts believe cannabidiol can have an advantageous impact on the mind, provide treatment, and decrease the symptoms of stress, major depression, or anxiety.

Home Cures For Anxiety

Do you realize there are many natural ways to lessen anxiety?

Simple methods like meditation, yoga, or deep breathing exercises can alleviate the symptoms of panic and significantly improve your current sense of well-being. As well as the best benefit is that you can practice them at home.

CBD Essential oil for Anxiety

Because the Internet is bristling with instructional videos about how to meditate, do a little of light yoga, or practice a breathing exercise, it might be a shame never to test it out.

While these strategies should in no way replace common treatments, they are doing provide a little extra help for those folks dealing with anxiety.

CHAPTER 2

What's CBD?

Even though cannabis plant contains tens of substances, only two have caught the interest of researchers - THC and CBD.

CBD means cannabidiol, the central cannabinoid within cannabis and commercial hemp. Years of research have figured, in a moderate dosage, CBD can have numerous health advantages.

Unlike THC (tetrahydrocannabinol), cannabidiol doesn't have any psychoactive effects on the mind therefore you can't get 'high' even though you get the dosage incorrect.

By having a primary effect on CB1 receptors, CBD make a difference on your serotonin levels - the neurotransmitter that takes on an enormous role in feeling disorders. And that's why using CBD essential oil for anxiousness, stress, and melancholy might end up being a practical option.

Whether it will come in the proper execution of pills, hemp essential oil, vape oil, real CBD essential oil, CBD gummies, or other edibles, cannabidiol could just be the next discovery in mental health.

CBD Essential oil Effects

If you're coping with a psychological problem like anxiety and discover it's extremely difficult to tolerate the unpleasant aspect ramifications of psychiatric medication, hemp essential oil or any other cannabis product which has CBD may be considered a good alternative.

As 2010 paper posted in the Journal of Psychopharmacology suggests, both human being and pet studies indicate that cannabidiol, one of the principal compounds within cannabis and marijuana, can serve as a practical treatment for anxiety.

Actually, some medical trials have revealed that CBD oil works well for an array of conditions, from chronic pain, multiple sclerosis, and Alzheimer's disease to depression, stress, and anxiety. Furthermore, it can lower blood circulation pressure and become an all-natural anti-inflammatory supplement.

But CBD essential oil can be of great use even for those folks working with specific types of anxiety. For instance, a 2011 research released in Neuropsychopharmacology, exposed that cannabidiol could reduce interpersonal stress and anxiety - one of the most typical types of nervousness disorders.

The bottom line is, cannabidiol appears to have important health advantages which will make it a complement worth considering.

Will CBD Get You High?

A lot of you are most likely worried that using CBD essential oil for anxiety are certain to get you 'high.' Thankfully, it's impossible to get 'high' off CBD, since it doesn't have any psychoactive results on the mind.

While cannabidiol will interfere with a variety of neurotransmitter receptors, the 'high' is always the consequence of THC. In a nutshell, no THC, no high.

And if that isn't enough to convince you, be aware that CBD essential oil, hemp essential oil, and other cannabidiol-based products can be purchased in most countries round the world - even countries where both medical and recreational cannabis is illegal.

Is CBD Addictive?

The simplest response to this question is no. The usage of cannabidiol, unlike the utilization of weed or cannabis products, isn't addictive.

As one record issued by the World Health Corporation in 2018 obviously states, at the moment no public health issues (e.g., driving while impaired of drugs instances, comorbidities) have been from the use of genuine CBD.

Even though you have a higher dose, there's no chance you could ever get dependent on cannabidiol.

Is CBD Safe?

Since cannabidiol proves to work in treating several medical and psychological conditions, we can reasonably assume CBD-based products are safe. Actually, millions of individuals across the world are employing this product regularly, without experiencing any medical problems.

Furthermore, individual and pet studies have revealed that cannabidiol is a safe substance that can prevent colitis in mice and also become a chemo-preventive health supplement for cancer of the colon.

Overall, experts can concur that using CBD essential oil for anxiety is a flawlessly safe and surprisingly effective option to traditional treatments.

Cbd Unwanted Effects

But because using CBD essential oil for stress or any other conditions is safe, doesn't mean there aren't any part results associated with this dietary supplement.

Much like any material that inhibits the normal working of the mind and body, CBD can involve some relatively unpleasant aspect effects.

The most frequent ones are:

- Dry mouth.

- Changes in appetite.

- Low blood circulation pressure.

- Dizziness.

- Changes in mood.

- Nausea.

- Vomiting.

- Diarrhea.

- Vomiting.

How Do Cbd Oil Assist With Anxiety?

Many anxiety sufferers use hemp oil, CBD oil, and other similar supplements to sleep better and offer with the continuous blast of worry-filled thoughts associated with this psychological problem.

One case survey published from the Permanente Journal in 2016 revealed that cannabidiol might be considered a safe treatment for the anxiety and insomnia associated with post-traumatic stress disorder.

CBD Essential oil for Anxiety

But that's not absolutely all. A 2013 research conducted on mice uncovered that repeated administration of CBD could result in hippocampal neurogenesis (the regeneration of neurons in the hippocampus).

For anybody who don't know, the hippocampus is a brain region accountable for cognition and memory space. Regarding depression or panic victims, this brain area is commonly smaller, which might explain a few of the irrational considering patterns specific to these conditions.

Overall, it appears cannabidiol is not only a placebo, but a substance that can directly impact the working of varied brain areas and neurotransmitter receptors.

Neurochemistry and CBD

Did you know the body has an integral system designed explicitly for control of cannabinoids?

It's called 'the endocannabinoid system' and it has a significant role in homeostasis - the body's capability to maintain a wholesome inner balance. This implies CBD essential oil and other produced supplements can regulate rest, appetite, mood, and therefore the amount of anxiety we may be experiencing at confirmed moment.

But what's truly fascinating is that substance can have a genuine impact on the mind receptors mixed up in body's response to anxiety.

For instance, cannabidiol works exactly like SSRIs (selective serotonin reuptake inhibitors) - a group of psychiatric drugs specifically created for anxiety and depression. By getting your serotonin on track levels, CBD-based supplements can significantly decrease the symptoms of anxiousness.

CBD Oil Dose For Anxiety

As the FDA does approve CBD as a product for an array of medical and psychological conditions, it's also firmly against companies that market cannabidiol as a '100% guaranteed' cure against severe medical ailments.

But even though CBD essential oil for anxiety can be an FDA-approved treatment, experts still find it hard to determine the right dose. You need to bear in mind that discovering the right dosage depends on several factors, from weight and age group to the strength of the symptoms.

Luckily, since cannabidiol is a safe and non-psychoactive compound, you don't have to worry about going for a dangerously high dose. The most severe thing that can occur is feeling a little nauseous or dizzy, symptoms which show you need to lessen the dose.

In addition, you can determine the perfect dosage for you by consulting with a medical expert or asking the maker.

Important Thing: Can CBD Help Treat Anxiety?

Overall, products produced from weed, cannabis, or hemp herb (ex-lover. CBD essential oil, hemp essential oil, cannabis oil, and so many more) can be a highly effective treatment for disposition disorders like stress and anxiety and depression.

Actually, a 2007 research posted by Chemistry & Biodiversity highlights, the plethora of positive pharmacological effects noticed with CBD get this to compound an extremely attractive therapeutic entity.

But merely to be 100% safe, be sure to consult your physician prior to deciding to use CBD essential oil for nervousness or any other condition.

chapter 3

CBD Essential oil FOR CHILDREN WITH ANXIETY

Are you worried that your anxious child may need medication? Well, stress in children is a significant problem, you aren't alone. Regarding to psychom, panic in children can be considered a stage in children's development as they start to connect to the world. However, some children have an elevated sense of dread and dread.

When such emotions of anxiety become dominant in your child's thoughts, then you should surely find a remedy. Among the remedies for anxiousness in children that is continuing to grow popular through the years is the utilization of CBD hemp essential oil drops. The best question is, is real CBD essential oil effective in combating stress and anxiety in children? Well let's find out:

- CBD drops

1. CBD essential oil maintains a reduction in anxiety

A report on the symptoms of a 10-12 months old who was simply sexually assaulted at five years showed that using cannabidiol products helped reduce anxiety steadily. The 10-year-old have been on pharmaceutical medications; however, the drugs were no more effective in combating the nervousness. She experienced and also suffered undesirable side ramifications of the drugs that led her to insomnia. The research discovered that the pharmaceutical medications only offered incomplete alleviation, but soon the stress kicked in and the kid lost control of her feelings. When she was launched to CBD essential oil, the anxiety decreased significantly, and the alleviation was longer-lasting than when pharmaceutical medications were used.

2. Improves the grade of rest in children with anxiety

CBD hemp products have a relaxing effect on panic and help enhance the quality of rest. The CBD receptors are situated in the mind and modulate the neurotransmitter release, therefore, helps prevent extreme neuronal activity. CBD consumables can lessen insomnia and enhance the quality of rest. Due to results on soothing nerves, many people buy CBD hemp essential oil online in reducing insomnia triggered by anxiousness, and chronic ailments.

In the analysis mentioned previously about the 10-year old who was simply fighting sleep after 3 years carrying out a traumatic experience, when she got introduced to CBD oil, it improved the grade of her sleep. Pure CBD essential oil drops, when found in a small dosage, enhances alertness and activity throughout the day hence escalates the strength of rest and enhances the regularity of sleep-wake routine. CBD works in enhancing sleep due to its anxiolytic and sleep-inducing results.

3. CBD does not have any adverse part effects

Among the explanations why many people are looking for alternate treatment for stress is due to the medial side ramifications of pharmaceutical drugs used for stress and anxiety. Most patients have tolerance issues with pharmaceutical drugs after long term use. When you get hemp CBD products for the treating nervousness, you have a definite conscience that your son or daughter won't suffer devastating drug-related results.

Contrary to what folks believe that using CBD drops for medication can cause dependence on marijuana, CBD doesn't have the psychotic activities of THC. It, therefore, cannot produce the euphoric results on the individual, neither can the kid be dependent on it. CBD is the element of cannabis that is accountable for its therapeutic potency; it does not have any recreational properties. Research done on mouse discovered that CBD in small dosages has no aspect effect.

4. CBD essential oil has a feel-good influence on the patient

Panic in children is mainly caused after a traumatic experience. Whenever a child has stress, he manages to lose the glamour of life as he's constantly enveloped in dread. The traumatic encounters of days gone by could keep playing in his mind's eye.

When CBD pills are found in his treatment, they work by inhibiting neurotransmitter activity in the mind and provides a lasting calming effect to the individual. During this time period, the brain requires a break from focusing on stress, and the individual can feel great about themselves. Research shows that using CBD essential oil for panic also boosts self-esteem in children experiencing anxiety.

- CBD tinctures

5. CBD essential oil can be utilized in changeover from

pharmaceutical drugs to coping mechanisms

Nobody wants to be stuck on drugs. Pharmaceutical drugs aren't only an encumbrance in costs, but can also cause dependency. Using CBD as a transitional medication from pharmaceutical drugs helps the individual adjust to coping systems like yoga exercises, exercise, and mediation. The consequences of CBD essential oil last long, so the brain is poised to be receptive to new coping routines.

The usage of CBD in medication keeps growing, because of its potency. The above-mentioned researches show how CBD essential oil can be utilized in treating anxiety successfully. If you want to buy CBD drops for an identical case, please get in touch with Terra-Vida online.

They have an array of CBD products to help you with various conditions.

Chapter 4

Using CBD Essential oil for Treating Anxiety

Lately, cannabidiol (CBD) oil has turned into a widely favored treatment for anxiety. Although some individuals take CBD essential oil to soothe their everyday concerns, others utilize it to treat much more serious conditions like generalized panic.

A compound within the marijuana herb, cannabidiol has increased in availability as cannabis use is legalized in increasingly more states in the united states. An increasing number of companies have started offering supplements, salves, and other products made out of CBD essential oil, typically touting these things as natural treatments for issues like stress and pain.

Unlike tetrahydrocannabinol (or THC, another chemical substance within marijuana), cannabidiol doesn't create a "high" when consumed. However, use of CBD essential oil isn't legal Atlanta divorce attorneys state.

Because of the legally murky character of marijuana, condition laws have a tendency to vary widely as it pertains to cannabis products of any sort. Therefore, it's important to discover if the utilization of CBD essential oil is legal in a state before using the product.

Uses

The most frequent mental illness in the U.S., panic disorders impact more than 18 percent of the populace each year, based on the Stress and Depression Association of America.

Although anxiety disorders are usually treated with

psychotherapy, medication, or a mixture of both, many people choose to forgo these standard approaches and self-treat with products like CBD oil. Relating to a study released in Cannabis and Cannabinoid Research in 2018, almost 62 percent of cannabidiol users reported that they used CBD to take care of a condition, with the very best three conditions being pain, anxiousness, and depression.

Due to too little research, researchers aren't sure how CBD essential oil will help treat issues like stress and anxiety. Some research shows that in addition to impacting the endocannabinoid system, cannabidiol may impact receptors mixed up in modulation of serotonin (a chemical substance messenger considered to are likely involved in nervousness regulation).

Cannabidiol (CBD) and its own Effects

Research

So far, the majority of the data for CBD's effects on anxiety originates from animal studies and lab experiments. For a written report released in the journal Neurotherapeutics in 2015, researchers analyzed this research and discovered that CBD essential oil shows guarantee in the acute treatment of conditions like generalized panic, panic disorder, interpersonal panic, obsessive-compulsive disorder, and post-traumatic stress disorder.

While there's currently too little large-scale clinical tests testing the utilization of CBD essential oil in the treating anxiety, a little research published in Neuropsychopharmacology in 2011 determined that CBD can help alleviate sociable anxiety.

For this research, 24 people who have social panic received either 600 milligrams (mg) of CBD or a placebo one hour. 5 before carrying out a simulated presenting and public speaking test. Additionally, 12 other folks with social panic performed the same test without getting any CBD treatment. Results exposed that pre-treatment with CBD significantly reduced stress, cognitive impairment, and pain while individuals were providing their speech.

The anxiety-reducing after-effect of CBD may follow a bell-shaped dose-response curve, suggests a report published in Frontiers in Pharmacology. After administering different dosages of CBD before a presentation and public speaking test, experts discovered that subjective panic measures were reduced with the 300 mg CBD dosage, however, not with the 100 or 900 mg CBD dosages.

Another study, posted in the Journal of Psychopharmacology in 2018, tested the consequences of cannabidiol in people who have high paranoid characteristics and discovered that cannabidiol had no effect on anxiety, cortisol levels, heartrate, systolic blood circulation pressure (the very best quantity in a blood circulation pressure reading), and persecutory ideation.

Cannabidiol didn't reduce reactions to negative emotional stimuli or reduce anxiousness in healthy individuals, according to a report published in Cannabis and Cannabinoid Research in 2017. Experts tested individuals' replies to negative images or words and intimidating psychological faces and level of sensitivity to sociable rejection after taking dental cannabidiol.

CHAPTER 5

Safety

Using CBD oil could cause lots of side results, including stress and anxiety. Some research shows that CBD essential oil may also result in the following part effects:

- Changes in appetite.

- Changes in mood.

- Diarrhea.

- Dizziness.

- Drowsiness.

- Dry mouth.

- Nausea.

- Vomiting.

- Low blood circulation pressure

Cannabidiol has been found to slightly increase heartrate at a dosage of 900 mg.

Furthermore, there's some evidence that the utilization of CBD oil can lead to increased degrees of liver enzymes (a marker of liver harm).

CBD oil could also connect to several medications, including benzodiazepines, calcium mineral route blockers, antihistamines, plus some types of anti-epileptic drugs. If you're on some of these kinds of medications, consult your physician before using CBD essential oil.

It will also be noted that, because CBD essential oil is mainly unregulated, products may be incorrectly labeled. Compared to that end, a report released in the Journal of the American Medical Association in 2017 discovered that almost 70 percent of most CBD products sold online are mislabeled and a quantity of products include a significant amount of THC. Since THC can aggravate nervousness and make your pulse faster than normal, it's possible that using CBD essential oil which has THC might make your stress worse.

A study review discovered that in the treating certain types of refractory epilepsy, individuals used lower dosages when utilizing a CBD-rich extract in comparison to purified CBD products, and found undesirable effects were less regular in those using CBD-rich extracts.

7 Natural Remedies to lessen Anxiety

Although it's normal to feel anxious every once in a while, in the event that you feel anxious without reason and these worries persist to affect your day-to-day life, you might have generalized panic.

Symptoms of generalized panic can include restlessness, feeling tense or on advantage, irritability, impatience, or poor focus. People could also notice changes in their physical health such as headaches, jaw pain, muscle pressure, difficulty dropping or remaining asleep (insomnia), dried out mouth, fatigue, upper body tightness, indigestion, bloating, sweating, and headache.

Natural Treatments For Anxiety

Even though some research shows that certain natural treatments may offer benefits, it is critical to talk with your physician before using alternative remedies. Take into account that it will not be utilized as an alternative for standard treatment in treating any health. These are a few of the natural treatments that are being explored for panic.

1) Passionflower

The herb passionflower (Passiflora incarnata) has an extended history useful as a folk treatment for anxiety and insomnia.

Two studies involving a complete of 198 people examined the potency of passionflower for anxiousness. One research found passionflower to be much like benzodiazepine drugs. There is also improvement in job performance with passionflower and less drowsiness with passionflower weighed against the medication mexazolam, however, neither was statistically significant.

Unwanted effects of passionflower can include nausea, vomiting, drowsiness, and quick heartbeat. The security of passionflower in pregnant or nursing women, children, or people who have kidney or liver organ disease is not established. There were five case reviews in Norway of individuals becoming temporarily impaired mentally after utilizing a combination product made up of passionflower. It isn't known if the other elements in the product played a job.

Passionflower shouldn't be taken with sedatives unless under medical guidance. Passionflower may improve the after-effect of pentobarbital, a medication used for rest and seizure disorders.

2) Bodywork

Therapeutic massage, shiatsu, and other types of bodywork are trusted to decrease muscle stress, relieve stress, and improve rest. Get one of these variety of popular therapeutic massage styles.

3) Brain/Body Techniques

Mind/body deep breathing exercises, physical activity, yoga exercise, tai chi, self-hypnosis, yoga, and biofeedback are just some of the stress decrease techniques used for stress and anxiety. Try different techniques and determine which program you can adhere to with a frantic schedule. Several great options are diaphragmatic respiration, the rest response, and mindfulness deep breathing.

4) Valerian

The herb valerian (Valeriana officinalis) is most

beneficial called an herbal treatment for insomnia. Valerian is also found in patients with moderate anxiety, however the research assisting its use for stress is limited.

For example, analysts with the Cochrane Collaboration reviewed studies on valerian for anxiety. Only 1 research met their quality requirements. It had been a four-week research evaluating valerian, the medication diazepam (Valium), and a placebo in 36 people who have generalized panic. No statistically significant variations were found between your groups, perhaps because of the small size of the analysis.

Valerian is usually taken one hour before bedtime. It requires about 2-3 weeks to work and must not be used to get more than 90 days at the same time. Unwanted effects of valerian can include slight indigestion, headache, palpitations, and dizziness. Although valerian tea and liquid components can be found, most people do not like the smell of valerian and choose taking the capsule form.

Valerian must not be taken in numerous medications, especially the ones that depress the central nervous system, such as sedatives and antihistamines. Valerian must not be used with alcoholic beverages, before or after surgery, or by people who have liver disease. It will not be utilized before traveling or operating equipment. Consultation with a professional health practitioner is preferred.

5) Kava

Local to Polynesia, the herb kava (Piper methysticum) has been found to have anti-anxiety effects in humans.

America Food and Drug Administration (FDA), however, has issued an advisory to consumers about the threat of severe liver injury caused by the utilization of health supplements containing kava. Today, there are more than 25 reviews of serious undesirable effects from kava use far away, including four patients who required liver organ transplants.

6) Gamma-Aminobutyric Acidity (GABA)

GABA can be an amino acidity that may be likely involved in the physiology of nervousness. Some prescription medications for stress work by influencing GABA receptors in the mind. The amount to which orally ingested GABA supplements can reach the mind, however, is unfamiliar.

7) Aromatherapy

Herb essential oils can be put into baths, massage oil, or infusers. Essential natural oils that are used for panic and nervous pressure are bergamot, cypress, geranium, jasmine, lavender, melissa, neroli, increased sandalwood, and ylang-ylang. Lavender is the most typical and forms the bottom of many calming blends.

Other Remedies

- Picamilon.

- Theanine.

- B Complex.

- Yoga

CBD Essential Oil And Anxiety

Stressed, stressed, and depressed woman holds her hands

across her chest.

Cannabis may aid relaxation, which makes it a popular option treatment for anxiousness.

Much of the study on cannabis products has viewed the utilization of marijuana rather than at CBD essential oil as a stand-alone product.

Some studies have discovered that cannabis will help anxiety. Others claim that having panic is a risk factor for recreational weed use, or that using cannabis can make a person more susceptible to stress and anxiety.

People thinking about managing their anxiety with CBD essential oil should look exclusively at research on cannabidiol, not generalized studies of medical weed. Although there are fewer studies on cannabidiol specifically, the research is promising.

A little 2010 study discovered that cannabidiol could reduce symptoms of cultural anxiety in people who have public panic (SAD). Brain scans of individuals uncovered changes in blood circulation to the parts of the mind linked to emotions of anxiety.

In this research, cannabidiol not only made individuals feel better but also changed just how their brains are taking care of immediate anxiety.

A 2011 research also discovered that cannabidiol could reduce public anxiety. For the study, researchers appeared specifically at cannabidiol to take care of nervousness associated with presenting and public speaking.

Research published in 2014 discovered that CBD

essential oil had anti-anxiety and anti-depressant results in a pet model.

A 2015 analysis of previous studies figured CBD essential oil is a promising treatment for numerous kinds of anxiety, including sociable anxiety disorder, anxiety attacks, obsessive-compulsive disorder, generalized panic, and post-traumatic stress disorder.

The report cautioned, however, that data on long-term use of CBD oil is bound. While research highly factors to the role of cannabidiol in dealing with short-term stress, little is well known about its long-term results, or how it could be used as an extended treatment.

A 2016 research study explored whether cannabidiol could reduce symptoms of post-traumatic stress disorder (PTSD) and anxiety-provoked rest disorder in a kid with a brief history of trauma. Analysts discovered that cannabidiol reduced the child's panic and helped her rest.

Risks

Man smoking a cannabis joint.

Smoking cannabis may present more hazards to health than using CBD essential oil.

Research on the utilization of cannabis shows that it could have negative health results, particularly if smoked.

Research specifically on cannabidiol, however, has found few or no negative aspect effects. This implies CBD essential oil may be considered a good option for individuals who do not tolerate the medial side ramifications of other medications for anxiousness, including addiction.

Not all says in America have specifically legalized CBD oil, even though some have legalized it for only specific purposes.

A person should educate themselves about the potential dangers of buying or utilizing it. While CBD essential oil is not outlined on the Managed Substances Take action (CSA), a person should seek advice from their doctor before utilizing it to treat stress and anxiety.

Because CBD essential oil is not regulated as a treatment for nervousness, it is unclear what dose a person should use, or how frequently they ought to utilize it. A person should seek advice from a doctor that has experience with CBD essential oil to look for the right medication dosage for his or her needs.

CHAPTER 6

My Experience Using CBD Essential Oil For Kids And Adults With Anxiety

First, I am not really a medical expert. I am a mom who is suffering from anxiety every once in a while, raising a child with severe anxiousness. During the period of my life, I've learned the worthiness of research, requesting questions, and getting knowledge to make decisions which have a positive effect on our family members health and joy. Years back I first found out about CBD but didn't give it much thought at that time. However, during the period of the past six months, I've spent an unbelievable timeframe gaining a knowledge of CBD essential oil for kids and adults and exactly how it can favorably affect stress and anxiety. As always, talk with a medical expert before any treatment for nervousness. I will give out my knowledge and experience using CBD essential oil for kids and adults with stress in the expectations it leads to a much better starting place for you or your stressed child.

My child has severe panic, therefore, I originally sought information on CBD essential oil as cure option on her behalf. I too have problems with anxiousness but it will ebb and circulation as time passes rather than impact me on a regular basis. I should notice, if you ask me, there are two very unique aspects of stress and anxiety; the physical symptoms that nervousness causes and the mental stress that cycles through your brain. I will chat more concerning this after I clarify more about CBD essential oil.

What I Learned All About Cbd Essential Oil For Kids And Adults With Anxiety

Before choosing to use CBD oil for kids and adults as cure option for anxiety, I managed to get a spot to learn whenever you can about it, just like I would for just about any other medication or therapy option.

Over-the-counter and prescription drugs is controlled through the FDA, and therefore all medicines of 1 kind and dosage have the same amount of medicinal value and are manufactured just as CBD essential oil is not governed by the FDA (like the majority of natural treatments), and therefore power and quality may differ greatly. Continue to keep this at heart when requesting questions or purchasing CBD products.

So, What's Cbd Essential Oil And Could It Be Legal?

CBD means cannabidiol, which is among the many strands of the hemp flower. The very first thing that lots of people think (or presume), as users of my very own family do, is that CBD is cannabis. It isn't weed. It's been referred to as "marijuana's non-psychoactive cousin". THC is accountable for the mood-altering ramifications of cannabis, however, there is absolutely no THC in CBD essential oil. CBD essential oil is also not hemp essential oil or hemp seed.

CBD essential oil is legal in every 50 claims and bought from many natural food markets around the united states. While cannabis is legal using states in the US, it is controlled in similar ways to alcoholic beverages, meaning that you can't utilize it while generating, working, or under a certain age group. Because CBD essential oil will not contain any THC, it generally does not have these rules on its use.

CBD essential oil is refined in ways much like those of essential natural oils. After the refinement process is complete, it is examined for contaminants, poisons, and cannabinoid content. Again, since this isn't governed by the FDA, always be sure you are purchasing from an established company.

CHAPTER 7

How CBD Essential Oil For Kids And Adults Works

While CBD oil and cannabis are related, they connect to your body in completely different ways. CBD interacts with the endogenous system, a body-wide assortment of cell receptors that play a simple role in the function of the anxious and immune system systems. In the event that you suppose these receptors have hair on them, then your CBD is the main element that unlocks them. Once unlocked, these receptors can help you your body in a variety of ways. Research shows that CBD essential oil can help with epilepsy, arthritis, depressive disorder, anxiety, cancer, rest, and MS among numerous others.

You will find differing thoughts about how better to use CBD for maximum effectiveness. Some individuals believe taking it, as needed is most effective. However, after talking to several professionals, I came across that taking it frequently tends to boost the performance for overall panic. Additionally, there are numerous ways to "take" CBD. Typically, the most popular is the liquid drops that are positioned under the tongue or with food. There's also pills, creams, therapeutic massage essential oil, gummies, edibles, and areas.

My Experience Using CBD Essential Oil For Anxiety

After spending months exploring CBD, I purchased an extremely respected make of liquid CBD oil drops from an area natural grocer. It generally does not come cheap. Top quality, high power essential oil is expensive rather than included in any insurance policies. The CBD essential oil which i purchased was $120 for an one-ounce container. It lasted for approximately 4 weeks. I thought we would use the essential oil on myself prior to presenting my daughter check it out. Here's what I learned.

I should take note, I experience mental anxiousness sometimes, where my brain appears to get stuck on the "be concerned" that just is constantly on the routine through my brain. When this kind of stress and anxiety occurs, I hardly ever have physical symptoms. In such cases, I have found that positive considering and time will be the only ways to lessen the anxiety triggered by mental poison.

Additionally, I experience physical symptoms like a rapid heartrate, the shortcoming to catch my breath, and a standard "shaky" feeling. I rarely "think" which is anxious of these physical symptoms, as my brain is not working with any stressed thoughts or concerns, rather the anxiousness is just appearing in my own body with physical symptoms.

I began taking a little dosage of CBD essential oil every day with breakfast time and each evening before bed. This didn't give me any emotions of immediate comfort, however, I had not been taking it during occasions of high nervousness or during a panic attack. I had been taking it on the routine rather than as needed. Ahead of taking the CBD essential oil, I used to be experiencing physical symptoms of stress usually about once a day. I observed within the first couple of days of taking the CBD essential oil that the physical symptoms vanished. To check my theory that the CBD essential oil was effective, I halted taking it daily. My physical symptoms came back. Since November, I have already been taking a little dosage of CBD essential oil a few times daily and also have got very minimal, if any, physical panic symptoms. I've found this to be as effective for me personally as the prescription drugs I once required.

The CBD oil hasn't had any positive influence on the mental anxiety that strikes arbitrarily. Once a poor thought starts swirling, I have to sort out my skills to control that kind of anxiety. I will note that I've not acquired any extreme anxiousness or anxiety attacks since using CBD essential oil, so I cannot talk with its ability to control this kind of anxiety.

My Experience With CBD Essential Oil For Kids

Within a couple weeks of noticing that the CDB oil appeared to be helping my small anxiety, without side results, I started giving it to my daughter every morning before school. (My girl does take prescription drugs for her stress and anxiety.) My daughter's nervousness presents in a different way than my very own.

She really challenges with the mental facet of anxiety. Mental poison are running right through her brain a large area of the day with continuous "what ifs". During times of high stress, her body also reacts with emotions of nausea, shaking, headaches, etc. Unlike me, she rarely gets the physical symptoms in addition to the mental concerns. She also only calls for the CBD essential oil (by choice) on mornings where she has to visit school. My little girl prefers never to take it in the evenings or on the weekends because "she hates the flavor". I really do not pressure this on her behalf.

I cannot say that she or I've noticed an optimistic influence on her panic with the CBD essential oil.

Sooner or later, I'd like her to consider it more frequently to find out if it has a far more positive impact. I truthfully think her anxiousness with college is so high therefore habitual, that the CBD essential oil is not strong enough to overcome her mental stress and anxiety.

Overall, I am happy with the response I've needed to the CBD essential oil for anxiety which i purchased. I intend to continue utilizing it daily, as I think it is effective for my needs. My child will continue utilizing it so long as she demands it.

I encourage anyone experiencing anxiety to consider this option, just like you would consider some other medication, natural health supplement, or therapy. It's important to consult with a medical expert before beginning any kind of treatment.

Choosing the proper CBD Essential oil for

Anxiety

With so many choices, it can get overwhelming to choose the right CBD for dental care anxiety. Be sure to research your facts before buying CBD.

Not sure what things to ask? Below are a few of the questions I love to ask a merchant about my CBD essential oil:

Is your product created from 100% USDA-certified organic hemp? I only buy CBD from organic hemp. That way, you don't risk unneeded contact with pesticides.

Who produces your CBD product? Reputable retailers can provide the titles of their CBD manufacturers.

How much of the product is in fact CBD? Avoid fillers. Ensure you get something that is 100% CBD. Some brands put in a small amount of peppermint draw out or other flavors, but which should hardly dilute the CBD. You don't need to get a CBD essential oil with other essential natural oils in it or diluted with carrier natural oils.

Who performs your third-party screening? This is an integral question to ask. Because supplements can be made by any business, there are shoddy products on the marketplace. Require the name of the third-party screening company and execute a quick web search to verify it's a genuine tests facility.

What dose do you recommend? This will change by product. However, I love to use something that lists how much CBD essential oil I will be using. That presents me they're assured in the quantity of CBD in the merchandise and its strength.

Is your hemp produced in America? Sourcing issues with

CBD essential oil. You may pay extra for an US-grown variety; however, the requirements of American growers have a tendency to be higher (or at least clearer) than somewhere else.

Any kind of natural product bears risks, as the supplement industry isn't as tightly controlled as medications and food. Be mindful as you get CBD products.

I also prefer to find good reading user reviews. Viewing many people pleased with their results helps me know I'm getting something that works.

How to approach Dental Anxiety

I understand that lots of people steer clear of the dental professional because of nervousness and dread. Using CBD for dental care stress and anxiety is one smart way to lessen this burden, but there are many other options you can test.

First, go directly to the dental practitioner regularly. The much longer you decide to go between cleanings, the greater creative your brain can get in what might happen. Keeping accountable to maintain with appointments will desensitize you from stress about the dental professional.

Be sure you floss at home. A lot of my patients have concern with pain from flossing. In the event that you floss at home on a standard basis, you'll uncover the pain is negligible! I inform patients who haven't gone to the dental practitioner in a while to floss at least one teeth before they come directly into getting accustomed to the sensation.

Choose the best dentist for you. Feel absolve to ask as much questions as you will need and don't let anyone bully you. You have the right to any second views or dental professional transfers you want. In case your dentist enables you to feel guilty or stressed, it's probably time for you to look elsewhere.

When I've an individual with oral anxiety, I practice compassion, communication, and distraction.

Acknowledgments

The Glory of this book success goes to God Almighty and my beautiful Family, Fans, Readers & well-wishers, Customers and Friends for their endless support and encouragements.

www.ingramcontent.com/pod-product-compliance
Lightning Source LLC
Chambersburg PA
CBHW030941240526
45463CB00015B/883